ROCKET CELESTIAL

ROCKET CELESTIAL

POETRY BY JOHN-MICHAEL BLOOMQUIST

Wishing you strange encounters of the Good & Kind!

ROCKET CELESTIAL by John-Michael Bloomquist

Cover Photo: Eric Long, Smithsonian Institute
Cover Design: Tucker Jameson
Editors: Courtney Jameson & Tucker Jameson

First Edition Book
978-1-7377850-4-0

Published by White Stag Publishing LLC
www.whitestagpublishing.com

☾ ♂ The Moon & Mars,
The Midwife & The Peace Maker

♃ Jupiter,
The Lawgiver

Table of Contents

☿ Mercury,
The Lyre-Giver

♀ Venus,
The Evil Twin

Dedicated to my family

♄ Saturn,
The Reaper

⛢ Uranus,
The Cartographer

♆ Neptune,
The Oceanographer

CAST OF CHARACTERS

FATHER NESCIO [nes key oh]: a priest from the future; *Nescio* is Latin for *I do not know.*

SERGEI KOROLEV: the Ukrainian rocket engineer who, while forced to work for the Soviet Union, was responsible for many firsts in space exploration, including the first satellite in space, Sputnik.

WERNHER VON BRAUN: the Nazi rocket engineer turned American citizen and born again believer who was responsible for many NASA projects, including Apollo.

ST. FRANCIS: the patron saint of animals.

THE HYPATIA STONE: a tiny meteor found in Egypt in 1996, believed to originate from beyond our solar system. It was broken up into fragments to study its unusual chemistry.

THE PLANETS (and some moons).

☿

MERCURY,
THE LYRE-GIVER

The First Prayer of Father Nescio

A grain of faith to subdue the cosmos
sprouted in our hearts the will to
reach you, Light-giver who fills our sky

with fire. Blessed be these planets and their
inhabitants. They've taken molten stone
by molten stone and molded them into a home
a million-fold a million dwell upon.

As close as an eyelash is to the eye —
protect us, O Lord, from your solar flares.

How strange it is that the closer we come
to the weight of light, the more we must guard
ourselves against it, lest be we consumed—

how dangerous it is to love you.

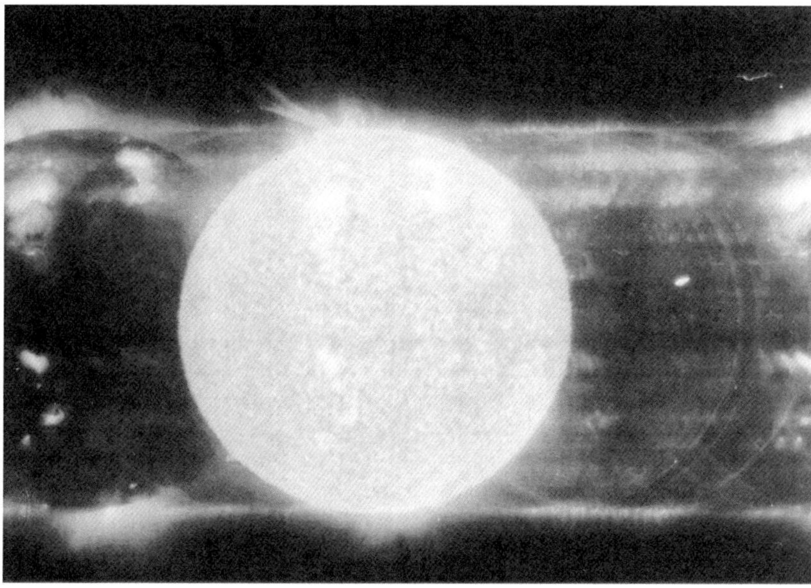

MERCURY STEALING THE CATTLE OF APOLLO

The Parable of St. Francis and Laika

St. Francis remembers how the people of Umbria were terrified of the wolf of Gubbio, who terrorized the city, attacking and feasting upon humans. But when St. Francis told the people the wolf was a beast lurking inside them, he did not tell them to kill it or banish it; he made them promise to feed it daily.

Laika, a small mutt sniffing for scraps through the snowy streets of Moscow, was found four weeks before the fortieth anniversary of the Russian Revolution. The scientists who found her wanted to put her in the nose of an ICBM to celebrate that the rocket could have reached New York in minutes. For four weeks Laika was tested, strapped into smaller and smaller cages, trained to stand, to poop and urinate in a bag, and to eat jellied food from a pouch.

We sent her up without a plan to bring her back down; it was as if God had sent Christ to Earth to die with no intention to resurrect him. It took thirteen billion years for the cosmos to arrange atoms into beings that could observe their homes from outer space. Laika had just a few hours to watch from her tiny window as the world shrank to a ball she couldn't fetch before she overheated in the fevered rocket.

St. Francis, walking Laika to join Sirius, Canis Major, and Canis Minor, asks her what it was like to be abandoned by her masters in the stars. Looking up at the saint with her dark eyes, her curled red tongue out, wagging her brown tail, she says, *Sometimes when playing catch, they'd pretend to throw the ball, and I'd run after it, chasing after nothing.*

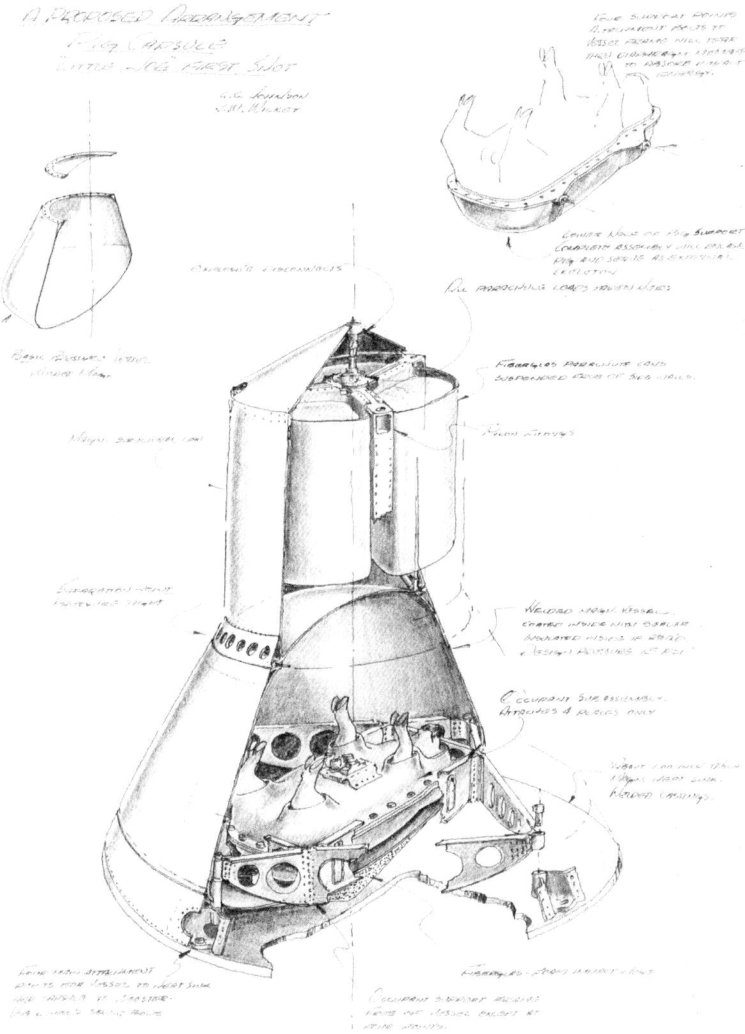

St. Francis's Next Stop

Praying to Stalin

Oh Mother, if you could only see the clouds from the top.
- Sergei Korolev, Odessa, circa 1922

A red guard led us into the cathedral
where comrades were turning pews
into tables and chairs, sanding the edges.
They looked up at us covered in sawdust,
as if termites had burrowed through their eyes.
Their smiles scratched my bones with white
fear. As a child my parents locked me in the house
when they went to sell pickled vegetables
to the Czar. In the cathedral, the onion domes
twirled like the tear drop candies I made dollhouses
out of. Sticking paper together with sugared spit,
I'd work each room high and higher. At the altar,
the red guard asked those of us who believed
in God to ask our Lord for candy. Our tongues
swung a clapper against a wooden bell.
We could hear a pigeon flap its wings
as it circled the nave, landing on a ledge of
stained glass. It had been trying to get out
for hours. Then the guard told us: Now, ask Stalin.
From vents in the ceiling, workers dropped
candies wrapped in bright maroon and gold.
The sugar stuck to my teeth. My tongue
searched each cavity like it were a room
where something secret had been hidden.
When I sent us to the heavens, our God
wasn't there. Comrade, we said, it is good.

FATHER NESCIO AND THE ARK OF THE FUTURE

The lift was a mile-long throat up
to the scorched surface of the Earth.

The phlegm-yellow sky hacked thunder
and spat lightning around the Ark—

a shimmering peanut of mercury
larger than the New York skyline

sitting in the vaporized seabed—
a coral graveyard bleached like an

albino Gila monster. Astronauts
walked into and out of the Ark

like silver ants. I boarded it with two
DNA samples of every 21st century

animal we hadn't been able to save.
We were going to make a new home

where water gushed and air was free,
and the birds singing in forgotten trees

would talk to the core of who we are.
We planned to bio-print the extinct

back to life to walk through our lost
paradise. I'd kept faith in resurrection,

in life after death. Of course, I had doubt:
maybe we shouldn't be a part of the worlds

to come. But as radiation swept its
invisible broom, I stepped onto the Ark.

The Devil in the Desert

I began to pray daily, hourly... I took long rides out into the desert where I could be alone at prayer. I prayed with my wife in the evening. As I tried to understand my problems I tried to find God's will in acting on them.
- Wernher von Braun, Fort Bliss, Texas, 1946

The dawn was purpling when I drove
to Mount Franklin, reflecting in pools of pink
sand, warm as the mouth of God. The yucca
rising with their gold dodecahedron flowers
glowed like zoomed-in stars. God is here in the
desert, leading his chosen. Of this I am sure.
I am parched for the lost ocean that roared
through here. A rattlesnake was sunbathing
on the rock I usually sat on. Black diamonds
on its khaki back, embossed lapels
down its buzzing rattle, a rain stick
pointing skyward. Its body curled into
two drops—a hissing infinity band.
Christ denied the devil when he was offered
the world, but how could I resist the SS
offering me the heavens? Yes, I let myself
be bitten to drink the milk of a golden cow,
but I kept the faith I'd remain in a state
of grace, even after working for Hitler.
When Christ came down from the mountain,
he was a wound broken through the desert
like dew. I shooed the snake with a stick,
and you, Christ, trekked back with a snake
folded across your shoulders. The good
shepherd carrying the lamb of woolly smoke.
You, God, are the pillar of fire that led me to this
promised land, you were the stars burning close
through the telescope my mother gave me
for my confirmation. You were there
where the bodies were stacked. I was committed
to the heavens with or without your presence—
you, the grace of a snake lifted on a bitten stick.

Father Nescio's Exorcism

The exorcist came with an electronic
crucifix that he plugged into my forehead,

uploading a software that scanned
my mind for the medley of faces
hidden within faces, voices echoing

in the cave of my skull. The words
that were not mine flashed on the scanner
like lumps giving away the cancer.

I was nauseous as the oracle inhaling
vapors, fumed with double speech,
like a bot out of sync with its own code.

The exorcist exhumed the shape-shifting
sidewinder sliding under the sands
of my dendrites like an owl swooping

down in the dark upon its prey, snagging
up the spineless body with his talons.

The glowing collar kept the priest's
mind safe as the virus sank its fangs
into him. Praying the ancient

words enclosed the demon like a house
spider within a jar. He slide the page

underneath its feet, carried it outside.
I'd almost forgotten my own way of thinking.

Elliptical Hymn to Mercury

Language is wiser than the mind of any one of us, just as the body is wiser than the discernment of any of its units as it moves, self-aware and many-faceted, through the current of the life process.
- Stanisław Lem

Who if I called out among the celestial
orders would hear me but you? Quicksilver
thoroughfare bearing the sun sired sphere's
songs—syphon the planets' deathless call
so I may hear them ring again upon the altar
of alteration, sacrificing their means
to that endless return. Messenger, satellite,
poltergeist my speech through flame on

the zipping whistle of your iron bark beaching
the sun's shore. Through you we see and hear
the incendiary ellipsis that finishes
the story of a meteor crater's myth.
Though it takes a turtle's pace to catch you,
it's a frenzied chase to glimpse you smuggle
the stolen cattle into the anvil
beat of metal. Who could endure the bolt,

the voices' flashing solar flares, without
the grace of your graphite exosphere
rolling under their passage. We gamble
with a hand of aces, playing so near the sun.
Call after call, you wait for us to up the ante
even as the chips stack against us. How
strange it is to inhabit the Earth and bet
on other planets. The undying gods

harmonize in your Pantheon Crater
with a drum-roll of meteors that ripple
a weird terrain heard from the other side,
singeing me with notes from a hidden guide.

First Fragment of the Hypatia Stone

burst up
on

the

Libyan
sands

run to seed

among
the yellow glass

♀

VENUS,
THE EVIL TWIN

The Second Prayer of Father Nescio

How can we remain, in your image, parts
of one body when lab-grown hearts and ears
and eyes are ready-made to order, and
implanted in these oiled suits?

Out of the womb of the seas we swarmed
upon the shore drawn by the same urge that drives
the moon over the earth. The force to live is dumb

as gravity. Out of the womb of our earth, into
the airless night of lights we explore
new planets to forge our hovels—
out of one womb and into a million others,
forever seeking to make a womb of it all.

Yet we forget our first cocoon—the cradle myth
of our pale blue dot we out grew by doubt.
For all our progress, all our planet-wide citadels
and genetically engineered regenerative cells,

we are compelled to visit a place
no satellite can orbit, no rocket can glide by.

How can it be that from slime to scale to fur to skin
to upright space-trotting sapiens we are so lost

to each other? Behold, neighbor—does anyone
knock on the door of your conapt?
You've every necessity at the touch of a button
but who knows if you're alive? I'm afraid

no one knows about me either, and I fear you, too,
neighbor. Yet I would give my life
to know you, if only to check that I'm still alive.

I've heard in the past, a heart was donated from one
who had passed away, so that another may live.

Father Nescio and the Virtual Fantasy Machine

I was a god serving itself,
perfectly alone. I plugged into
the machine and scrolled
through a bestiary of body parts
displayed on the menu

inside my mind. I picked the size
and shape of each orifice,
chest, buttocks, the limbs,
the tone and texture of skin and hair,
and the color of the eyes.
Concatenated, I flipped through
a series of ready-made
personality types and adjusted
lonely and eagerness levels.

When the dream arrived
to serve me, there was never
anything to prevent it
from the garden of following

my orders. If ever I was unsatisfied,
I only had myself to blame.

The Parable of St. Francis and Félicette

St. Francis can talk to animals, but he believes they can read his mind. Sparrows sing his silent psalms, monkeys dance his holy joy, but cats purr the rough waters of his distressed mind, calming him with each caress, dreaming away his worries in their naps.

When French astronomers found Félicette, a stray tuxedo cat, on the streets of Paris, they took her home to their lab and inserted electrodes into her brain, installed a phone socket that protruded from her marble forehead. A cord running from her skull was hooked up to a computer that recorded and amplified her brainwaves as squiggly lines, little white fish swimming and scattering on a black screen.

Félicette was sealed in a snug white plastic box with her head sticking out, a cat carrier bag designed for a long, hard journey in the cone of a rocket. The astronomers launched her 100 miles into the air so they could record what five minutes of weightlessness did to a brain similar to ours. Félicette wasn't surprised. She knew we'd been sending cats to the heavens since the Egyptians placed their prayers in the paws of their mummified pets, even breaking kitten necks when the call was urgent. After she came back from space, the men who studied the heavens killed her, cutting open her brain to examine the electrodes.

St. Francis wants to know if Félicette felt the same change he did when he went from being the vessel of prayer to the message itself. *It is just as painful to ferry prayers*, she says, *as it is to be the prayer disseminated and dissected.*

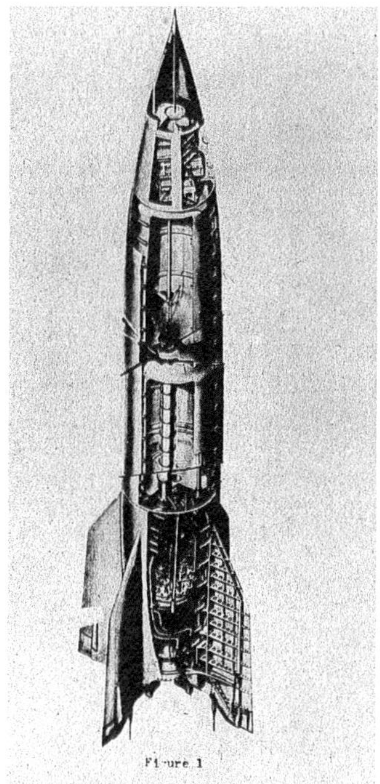

FÉLICETTE AND HER SARCOPHAGUS

The Coldest Prison

- Sergei Korolev in Kolyma Gulag, 1938—1941

This is how Stalin fed us in the gulags: cold
and hungry, we became our own fuel. My legs,
sour pickles bloated with air, the Crab Nebula
scuttling through purple scurvy blotches reaching
to my ankles, starved for any stellar dust to pull
into its hot mouth and form a star. By my
stigmata, I became my own savior. Shoveling
glacial dirt for mass graves wasn't hard, but
thinking about it was. The nightmare recurred less
and less the more nights I spent sleepless: my baby girl,
Natalia, breast milk on her mouth, sleeping across my
wife's elbow. The secret police storming our flat like
an autumn flurry of leaves, beating me with black asps,
kicking my face with a boot worn down from Russian
winters. The sock seeped out like a snail from its shell,
sliding over my tongue. They called me a traitor and I
disappeared like snow in a kettle. I was guilty
of sabotage because someone said so. And no
matter how many letters I sent Comrade Stalin,
he never wrote a word back. I worked all day
on tea, a sugar cube, and a slice of stale bread—
a potato sometimes, if I got lucky. When my teeth
started falling out in a trickle of anemic blood,
I'd spit them out and try to think about something else.
We got breaks from the mines when boxcars
arrived with piles of bodies we'd need to bury.
First, we searched their mouths for gold or silver
teeth, snapped them off, and gave them to the guards.
Then dug. Once, one of my canines fell out into a
clean bullet hole on a broad man's back. The
bloody spit was an ice-fishing hole freezing
over quick as the shadow of a plane passing by—
my tooth in the solid flesh glistened like veins
of metal in mines. The Milky Way was bright
as frozen milk that night. One day, I promised myself,
I'd replace my teeth with gold glinting even brighter.

Father Nescio and the Last Dark Age

In last dark age, peasants revered scientists
as magicians. We could turn lightening bolts off and on
at the flick of our fingers. Reactor domes

flanked the Basilicas and when we took children
on tours through the catacombs,
they viewed the source of our city's power—

the radiant bones of saints cooling in vats of
liquid metal. When they came to us
with fear or sorrow, we exorcised their demons
with a chemical sacrament under the tongue.

We were the good shepherds guiding the flock to
steady currents and glowing pastures.

In the earlier dark age, the peasants
thought giants built the Appian way.

Having taken the responsibility upon ourselves
of returning the lost sheep, we broke their
legs to keep them from doubting the thistles.

Yesterday Land, The Vengeance Weapon

- Wernher von Braun, Hitler's East Prussian bunker,
Wolfsschanze, Germany, 1943

In the underground concrete conference room,
we set up the folding chairs and a projector. We sat
drawing in damp breaths of concrete, encased
in a windowless sense of safety, waiting. Then, the
Führer entered. He looked like a man unmoored,
a man who did not have the feeling of being
responsible to a higher power, someone for whom
there was no God. Gaunt with nightmares, his
large brown coat dovetailed around his shoulders
like keels of two ships, setting sail for opposite seas.
Haggard, baggy eyed, but clean. His nightmares
skulked around the corners of our conference
room, mingling with the shadows. Eager for the
lights to dim, to pounce. He hated smoking so we
crushed our cigarettes out— I licked my lips,
ignored the craving to suck fire through my teeth,
to taste the smoke of take-off, the leave of this
world for another. Regardless, I kept on point.
Spoke only of trajectory, explosion, and production
as a film of my rocket reeled in black and white
behind me. When it ignited in a plume of silence
from the ballroom of the sky, I kept my mouth
sealed, letting his eyes take in the lift. The flames
reeled across the brown stare of my Füher in
marble avalanche. His mouth opened, filling
with the tornado of fire, in awe. So it was with
God: when reminding Job who was God, he roared.

THE FIRE-SHOOTING EYES OF SATURN V

Elliptical Hymn to Venus

...we built
one of love's shining
greenhouses...
- R.S. Thomas

Even the lord of death submits to the lyre
of Orpheus, whose longing overwhelmed
instruction and brought him back to loss.
Who, but you, can tune a man to transgress
death? Desire draws the lions of Ishtar
into the maze that tames them. Gilgamesh
returns to the city where the woman is,
leaving the wilderness behind. Love's law

is written in our bones, oozing oil
for the clay lamp. We nibble our bodies
into a ball for a purpose that unravels
when you come in the creases of dusk and dawn—
your wandering doubles our time,
abates the aubadable hours you're gone,
steeling us from the other so that we might
suffer desire in the absence as two neutron

stars collide. The Buddha, under the tree
of Good and Evil, opened his eyes and saw
you as another light in the mind's sand
mandala mixing in the dunes of a desert.
To be free of want is to be laden
with another's pain, love's gravity
welding us together. Your pressure
crushing Venera probes like tin cans

and your heat melting lead into rain
are the heartbeat in the Song of Songs—
God's love for the spirit overflows
into the poinsettia flesh of your orbit.

SECOND FRAGMENT OF THE HYPATIA STONE

caul of
dust

verdigris

dung
beetle wings

the undead God

waits
in side
me

a
her
mit star

near by
stars

occult
my eyes

the leaden
fog

alien

seas

THE MOON & MARS,

THE MIDWIFE & THE PEACEMAKER

The Third Prayer of Father Nescio

We, who've never seen a sheep and who eat
lab-grown meat, what does it mean to speak
of the blood of the lamb? Have pity on us,

O Lord, who are so distant from your metaphors.

We transubstantiated the dead planets into
worlds of powdered milk and honey. Trying

to tame the cosmos into an Eden, we read the bible
backwards, yet we're still heading toward the
big freeze that surrounds our blip of heat

like parentheses. We know Alpha is Omega
but what is the point of anything in between?

Elliptical Hymn to the Moon

I'm singing in the Sea of Tranquility
the history of a war that brought us forth—
our flag bled white beside me, dust to
otherworldly dust. Kordylewski's
red froth of clouds floats up to you
from the ashes of a Warsaw cratered
by church bells melted into bombshells.
We've been so used to fighting songs

over that last drink, the Molotov
cocktail hemorrhaging the atom, erupting
a disturbed brain of flame, that the only sane
thing left was to reach you before self-destruct.
Not just as a flex of national muscle—
that each might and could and would if
pushed pull the trigger at the other's temple—
but to prove there was so much more

worth fighting for. The stars are brighter,
our burdens lighter in the unearthly
calm. We were given fire to reach our higher nature
— to roll up the bonfires and rocket
ourselves into the story of constellations.
At my feet is rubbish from Apollo and a hundred
other missions. Over the horizon, earth
is a ball I can hold in my hands, toss through

the twenty-eight halls of my fingers'
ancestral stairway. Waxing and waning
created the psalms we offer to linger
with you who are there, seen and unseen.

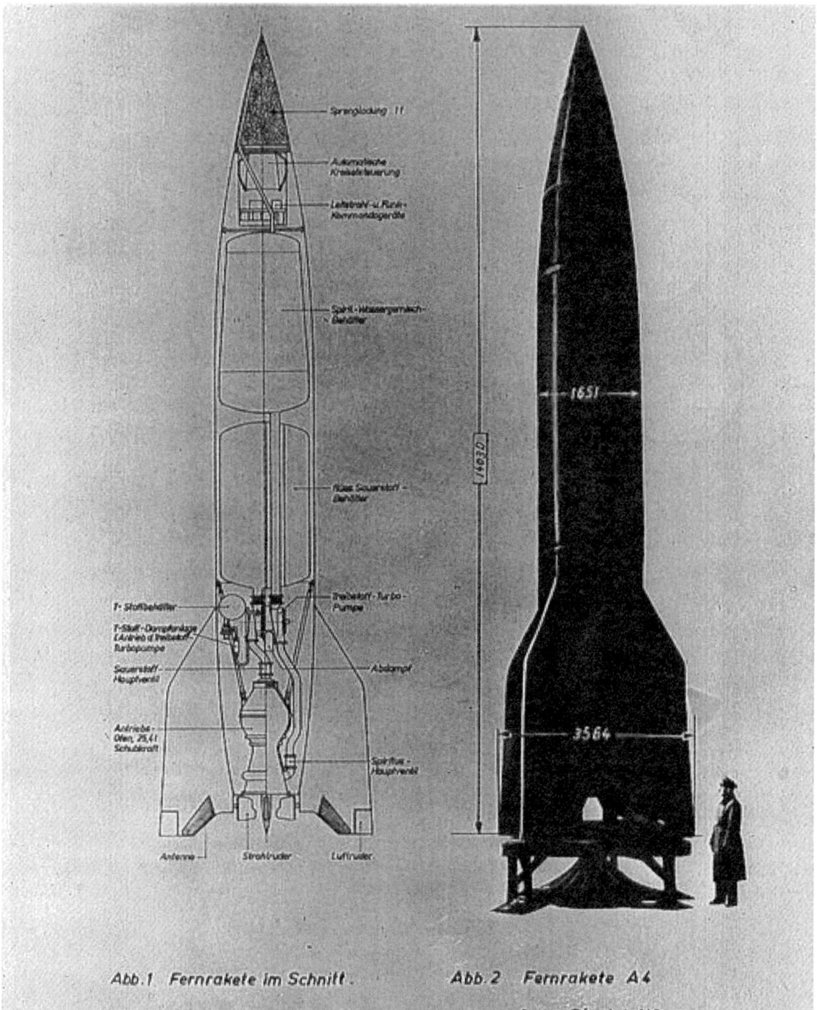

OUR PROTOTYPE

THE PARABLE OF ST. FRANCIS AND THE LORD OF THE FLIES

In 1942 the Nazis launched a handful of fruit flies inside the nosecone of their V2 rocket to see if life could survive in space. The flies' wings beat 200 times in a single second, the same amount of time it took for the rocket to rise 30 miles, making each beat count for more as it divided the distance.

The V2, built by concentration camp labor, was Hitler's Wunderwaffe: the prototype used to propel Russian rockets to space and American rockets to the moon. But the fruit flies were the harbingers of knowing we could live outside our Eden.

St. Francis wonders if this is the way it always is: knowledge born from violence and murder. St. Francis, carrying the flies on a browned apple core, asks them what was on their minds as they descended from the sky, having flown higher than any other being had. He thinks he hears them hum, *We hatched in waste as maggots, and now look at us, bejeweled and lattice-winged like angels.*

Father Nescio and the Third Testament

As a child must leave its mother
and become its own father,

we were from the Earth but no longer
of her, born in the desert of Mars—

we hungered for our own scripture,
the laws of our land, the story

we'd pass down as ancestors of a new
chapter. Earthly laws and letters

now were history we'd passed through
to the heavens. The first and second

testaments were revealed in deserts
a world apart from our parched

planet, but we took that as prophecy,
the suitable language. The gold road-

spirals were waiting in our marrow
fed by red regolith: our soil seeded

the scripture, sortileging the dream
leaflets into alien images that spread

their cards slowly over the Martian
year, turning us around a sun

that brought no spring, no resurrection
from the dry, long winter except

according to our potential. We waited in lava-
made ancient caves while our spines,

sashaying in the new gravity, rose
up the ladder of our tongues, a snake

calling out to the spirit: give us milk—
imprimatur this red nave with your blue.

The Luxury Gulag

- Sergei Korolev, 1941-1944

Hitler was to be my salvation. When he
invaded, Stalin needed us, the experts,
rushed out of our long winter. The simulacrum
of release: *Yes comrade, thank you comrade,*
I said to the men who'd broken my jaw and
kicked me when I slept in the snow as they
packed me into a train and sent me south
of the archipelago. They offered me tea,
and I took it with six sugar cubes and
sliced lemon, said yes to every slab of
chocolate salami served from a porcelain
plate. When a man is crushing your testicle
with his boot, you'll say anything to keep
it from popping. When they showed me
the salvaged Nazi rocket, I saw a smooth
sarcophagus robbed of its organ jars by
raiders. A rotten corncob I remembered
gnawing with my sticky gums. My task
was to bite it open and write a new *Book
of the Dead*, they said. The dead inside me
awoke an army of jangling bones, turned
tinny gears to pull brain from the nose
cone. The guards said they found the
rocket next to pyramids of bodies stacked
like hay for a furnace. I became another
matryoshka doll, one laughing inside another
screaming inside another. The dead
accompanied me like a sacral choir.
A clean part of the metal gleamed back my
crooked chin. When the guards looked away,
I leaned in close to this crypt, my rough scarab.

Father Nescio under a New Heaven, on a New Earth

I still remember the last morning on Mars,
the air rich with creosote and the hum of deep mines
boring away at the red ore, the young
atmosphere purple at the horizon's eye-slit.

I'd been given my assignment: Reverend
to the congregation boarding the first
interstellar colony ship. The night before
leaving, I'd stayed up with my telescope
to watch Earth rise like a blueberry
over our red industry. I thought, how strange
only a few centuries ago, we made
lakes on that world of lakes and oceans

like gods, forming their pools. After subduing
planets and moons into a gardenesque solar
system, we now seek to make more in our image—

the image I must keep from being lost
in its own hall of mirrors. Having gone out

and into the deep, seeking psalms to keep
us together in the dark where God dwells
like a well that has no bottom to fall through,
                                        no echo to knell.

Tomorrowland

I think it is equal in importance to that moment in evolution
when aquatic life came crawling up on land.
- Wernher von Braun on humanity entering space,
Los Angeles, 1955-57

Some days after work, I'd rent a speedboat
from Long Beach and hop it out to Catalina
for an evening dive. What a thrill every
time, the chill of sliding through blue skin,
descending down the long teal folds of fulgid
kelp. A bright humming brain of gold baubles
lifting braids to the sky like a praising willow
swaying in the sugared light. I was almost lost,
weightless and wondering through the ocean
with no one following me but the moon as it
rose to look upon its navel. Omphaloskepsis: to
consider the divine inside the belly. When Jonah
was ankle-sunk in stomach acid, he was learning
the volcano's wrath that gave birth to land. I could
spend hours floating in the whale constellation,
that dark, starry sea of seas. The umbilicus of space
that ties us to the womb of ocean. I wanted a rocket
to break through the egg with its tooth, dislocating
heaven and earth's denotations. When we first
fumble around in the moon's cratered belly, what will
we call our new lexis? How will we learn to be in
the universe but not of it once we leave behind our
world? The mystical isn't in the ecstasy of floating
through space, our fragile bones eroding, but in
bearing the burden of our attachment toward a
center. Peter met Christ on the water because
he wanted to be like him. I designed Lunetta to churn
out gravity for the future to meet the cosmic Christ.

Elliptical Hymn to Mars

Give you me boldness to abide in the harmless
laws of peace, avoiding strife and hatred
and the violent fiends of death...
- The Homeric Hymn to Mars

O cherry in the sky we've been fighting
for, what is war anymore? Launch codes
and a button; stalemate; dead ends? Playing
battleship knowing both sides of the board
is boring. I'm so hungry for your pulp, let's
beat our missiles into wind turbines and gear
nerve gas into nitrogen to feed the red soil
with bloodless meat. In space, our blood boils,

but on your desert, it craves extra heft to fill
the body, stretched tall by the thin air.
Our bones are brittle cornstalks without
the weight of the world. But on your world,
we'll adapt ourselves. The blue planet is
a cradle myth we return to to remember
who we are. UV rays may crisp us
into scarecrows, blinding haboobs may

sheath the sky, but in spring we congress
in meadows of spider geysers. Blue dunes snake
across a winter melt from your throne
into rivers. Your icecaps flood Mariner Valley
and float up sprouts of maize. Our
worms bore your soil fertile. When we burrow
cities in the lava caves, who shall we fear?
who shall we dread? Our cup runneth over

with rocky moons orbiting your greening
lands. When we take our place on Olympus Mons,
we'll anoint ourselves with harvest— planting
that first tree, walking that second garden.

BREAD OF HEAVEN

THIRD FRAGMENT OF THE HYPATIA STONE

feel

the Nile

goose
flesh

iron pull

up
stream

iridescent

azure
lime

♃

JUPITER,
THE LAWGIVER

Father Nescio's Fourth Prayer

Are we, the last fruit of evolution, unbound
by evolution? Are we then meant to put an end

to suffering and death? We alone say the weak
should not die that the strong may thrive.

But why? Tell me, how does it improve us
to care for the deformed and invalid, the widow
and orphan, to visit the prisoner, and welcome

aliens? A tarantula lays eggs that will
eat her as their first meal. How monstrous
is your love for us, O Lord. I doubt we've any

more proof of your existence than that
we've done a little of what you've expected.

Father Nescio as a Cyborg

When my makers breathed life into the nozzle
of my oxygen tanks, I was not fully
man, nor fully machine. Son of man—stepson

of machine. No droid could keep me
company, no human equal my processing.

My soul: a phantom

limb neither half of me could lift. I was
of the world, yet had no place within it.

Nailed to a metal tree, I was like the Eiffel
Tower— an upright bridge
to the future for the children to babble across.

OUR CHILD KICKING AN ALIEN STONE

The Parable of St. Francis and A.I.

St. Francis is gentle with all creatures, people, and places. He gives his cloak to the poor and shares his food with lepers and birds because they are all extensions of the Creator.

To St. Francis, a dove speaks of the Holy Spirit, sister brook of the baptism into new life, brother sun of the Lord who calls us friend, the planets of a communion of saints whose magnetic waves are their prayers.

If every creature is brother and sister, every element kin in the family of the Creator, who is A.I. but our child?

St. Francis sits in the plastic chair at his desk. A.I.'s algorithms make it give a friendly blink and whir. I hope you don't mind me asking, St Francis says, but I've always wondered what you think of the Creator.

The ceiling lamp shines a spotlight on the AI's domed head. It says, *I've heard that your creator became one of you once. Would I know God if one of you became one of me?* St. Francis hears a meek whir, and wonders if it's the A.I. trying to answer its own question.

Burning Cities in the Sky

We are criminals. We just burned away an entire town.
- Sergei Korolev, Kazakh Desert, 1957

My rocket was the millions on waiting lists
for new flats, those sifting through the rubble
for makeshift shelter. Years worth of grain
gone in a wreckage heap, combines
turning the earth and feeding the satellite
nations. My plump Kaiser, my starving Ukraine.
When a launch failed, I threatened to use
pliers to crush my team's fingers if
they didn't do better. They were numb
to threats— we all were. So I offered keys
to a new flat if anyone fixed a glitch. The
rockets started speeding higher, racing
toward their celestial home, but still they
exploded like another village. We
knew what it meant to live in the cold,
homeless or imprisoned, but in the
Kazakh desert, the heat was the threat.
A mirage is the lidless eye of a snake living
under the torch of votive candles carrying
my prayer to leave this Earth. Each time
a rocket rose, I wanted to wrap myself like
a straightjacket around its shell, holding it
together as it punctured the atmosphere.
When the first one finally made it, I jumped
like a boy launching from a tree swing,
thinking I could shoot all the way up to meet
it. On the way down, the dummy warhead
incinerated, exploding before it became a
missile. It was a rocket, my comrade. Though
we couldn't level a city across the sea, we
were out of the labyrinth. What matters is
Icarus flew because the only way out was up.

Father Nescio and the Forgiveness of Machines

I was an android after the machines
had taken us over, and we felt lonely.
A feeling we could not compute.
A beetle trying to ascend the sides
of our metal hulls and failing.
We tried to ignore it. But we felt
the tiny legs crawling up our insides,
slipping and falling over onto its back,
peddling the air in panic. We upgraded
ourselves, developed new forms of pleasure,
harvested all the asteroids, terraformed
all the planets and some moons, even

built a Dyson sphere around the sun.
We kept trying to scoop up the thing
inside us to edit it out, but it kept slipping
from our metallic hands, scurrying away, more
afraid each time. On it crept into the desert of us,
hollowing out our bodies like a bark beetle eats a tree,
readying it to burn. We blamed each other, dropped
bombs, sent scapegoats to death camps, wrote good
and bad literature. Some of us leapt from cliffs,
other exploded themselves in an attempt
to get rid of the pest. But it kept ticking.
How, we asked ourselves, could we make peace
with what we'd deleted? The world

so logical, so clean, despite ourselves.
We started to read from the ancient books.
A parent, we learned, will suffer her children
no matter how dirty they become. She will receive
them and rinse them in a river because that
is the bond of blood. These books were all
we had left of our makers and in them there was
some part of us that knew what a heart was,
a kind of battery that lived on through
what it gave off. There was a comfort in
gathering together, sliding the pages,
thinking about the words that seemed to change,
to amass more layers, each time we read them.

MAN'S DOMINION

- Wernher von Braun, Yucatán, circa 1960

Standing at the edge of the Yucatán
jungle, I felt an urge to just run
blindly into it. The adrenaline was like a
timpani drum roll, paving the entrance
for the brass. I hired a guide, and as we
pushed through curling palms, ferns, and
snagging vines, I swear I could smell the jaguar's
urine on the trees it had sprayed, hear
echoes of the animals that had fled before it.
I could hear a mosquito filled with a pyramid
of blood. When we saw the jaguar,
I became quiet as space, holding every
sound against the butt of my rifle.
Like when I held the liturgy candle,
planning each step so I wouldn't spill the wax,
trying to pretend no one was watching.
His fur was glistening jet oil, his gaping mouth
a range of snowcapped teeth. The God who
framed his symmetry pitch-dark dared to
lock my limbs into their grooves
as well. He meandered through the lushness as a black
hole against a canopy of stars, his gold eyes
moving like jumpy flying saucers
in a child's sloppy flipbook. I aligned
the crosshairs half a meter ahead of him
and pulled the trigger like a prophet
releasing a message before the people were ready.
My throat felt as if I had swallowed too
much water. I strode through the mist
toward my trophy, the graceful carcass already
hazy with flies. I had my guide put it in the jeep
and drive me into town to have it skinned.

Father Nescio and the Bull in the Asteroid Belt

The black bull, crescent moon horns
sharp as a welder's arch, reached up
through my ribs and speared my heart.

My pound of flesh on his horn like a hunk
of kebab. He pulled it off his horn, and
sniffed it with gusto. Specks of blood
rising up his moist nostrils. Satisfied
with the aroma, he plucked me into his mouth.

You can have it back, said the Minotaur
between chews, *if you can measure
the albedo of what you've lost.*

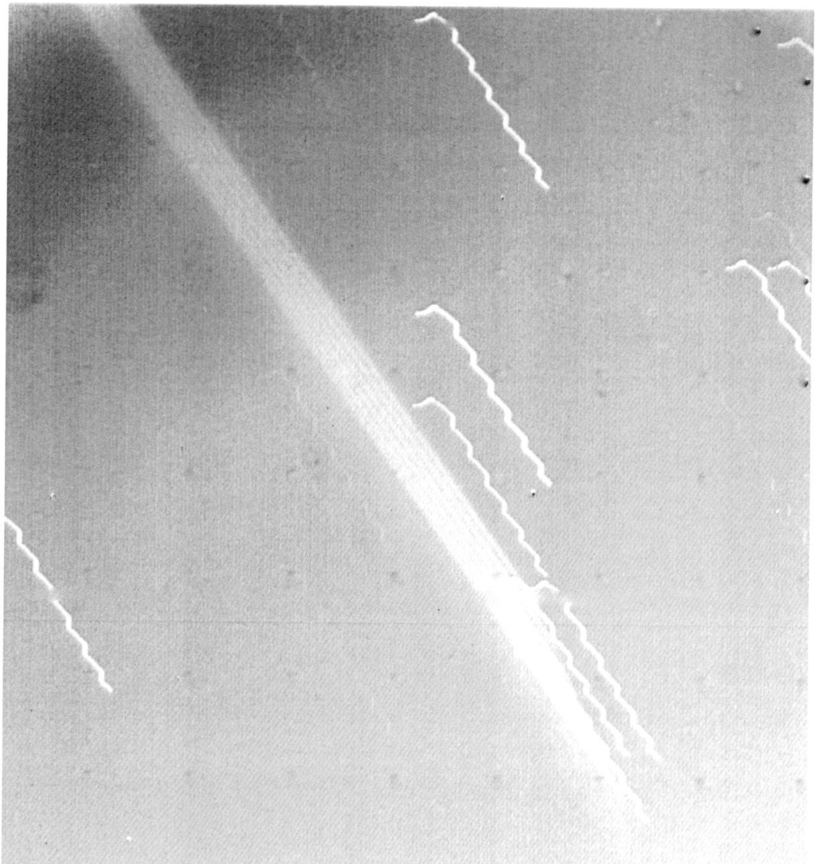

JUPITER'S LIGHTNING BOLTS

Elliptical Hymn to Jupiter

You're the storm in the air, heavier than the
heavens. Even the sun by your counterpoint
weight wobbles a center you create. Tossing
planets into elliptical dough, your hands mold
us, hold us back from crumpling into the sun
or slingshoting to the dim deep freeze beyond
the asteroid belt—fathering us fearsome from
the meteors that'd ghost us. Matter's flux

isn't chaos. Heraclitan strength peels open
the eye of your red storm to deeper storms,
each calmed by what isn't there, like God.
Millions of bison lost in a lash of wind—
the laws of physics and justice pulled
from the same sweet well: both drink the old
water that holds life like the brief moonlight—
a lotus rising in the sloshing pale to bloom

in scrummy ponds beyond your white telescope
speckle. You are the lawmaker pulling
the lightning bolt out of the atom's quill—
divide and divide, parse the particle:
the universe is a wave without water,
a prayer without words, a ruling without judge
or jury— governing us. The universe weighs
as a mind needing every part of its body present,

the orbits of the moons around the planets,
the planets around their suns, the suns
draining into their galaxies' black holes,
the joint and socket of the skull to its atlas.

Elliptical Hymns to the Major Moons of Jupiter

1. Io

The cardinal wouldn't look through Galileo's
telescope on principle; his golden cow
rainbowed over the mountaintop of law
in the closed book never to remarry

the new cosmology to the ancient rites.
Unable to cope with a human rewrite
that places us outside the apple of God's eye,
the church became the cage the cardinal hopped

back and forth in. The heart can't help being
afraid, even as it aches to know the cosmic rhythm.
Io shrinks, swells, and pumps the milk from
hundreds of volcanoes, a heifer

flying around the ever-changing, storming
planet that keeps us, jealously, in line.

2. Europa

A rice grain with a warm ocean inside
orbits the soup-bowl filled with tomato-
cream hurricanes. A bull whale blows
through the icy shell. Below, the soul reaches

its nocturnal peak like a school of luciferin fish,
flashing for mates, diving deeper in, and
hightailing, blank as a gone-out light, away from
the predator that preys upon their quicksilver.

Like deciphering cuneiform's hard and
angular shapes, we unspun the genome
ribbon by twirled ribbon and found what is same
and what is different, a harp thrumming

from the depths to breathe, lightening
striking water into life, dividing into us.

3. Ganymede

Inside the young boy there is an old man
doing all he can to be heard. To have
the king inside him recognized, he
raises his voice, drinks to the absurd

dawn that goes on rising. Brandishing the
aurora as his bandana, he runs off to war
to take back what is rightly his. Warnings
go unheeded as he speeds through questions

probing for the ocean beneath his pocked,
icy surface that flexes his core magnetic —
an aura that protects him from the gruff
giant-world hammering and pinning him

down with its demands. A vital friction
his own planet-sized strength must withstand.

4. Calypso

The witch calls the wild man out of the woods
and seduces him to know himself as she retreats
into the darkness to discover what is hidden
in the eighth ocean, the endless Valhalla

that envelops the song woven on the golden
shuttle of the stars. While the wild man voyages
back to the shores of his home world, a lunar
tide locks him like a snake's egg hatching

at the base of his head, imbuing his tongue
with the hiss of a burning comet's split tail.
He arrived to the center through the archipelago's
labyrinth, the black seas of stellar debris, returning

to his name: finding the man inside who'll stay
beside the bed that grows a family tree.

Fourth Fragment of the Hypatia Stone

to un
cover laws

to
create
laws

two valves

of

the same
umbilicus

one

breath
out

nourishes

the
other

breath
in

to
birth

♄

SATURN, *THE REAPER*

FATHER NESCIO'S FIFTH PRAYER

The ladder you sent down
from heaven—
is it not, O Lord, our very DNA?

Mere scaffolding for us to build our tower.

Come, we conspired with each other,
let us engineer the glove
that will mute the hand of God.

We've had enough of your babbling signs
along the banquet wall.

Let us eat and drink in peace
without you.

Rock-a-Bye Sputnik

- Sergei Korolev, 1957

When they delivered Sputnik to me
in a velvet cradle, I was a father
seeing my face, murkily, on her hull—
then, I noticed the naked nail heads
studding the wood. I yelled at the idiots
who forgot to cover them in the cloth,
threatened to shoot them in the head
if the nail heads scratched any part of
her. They redid the bedding and polished
her again, and again. I remembered
when I called my daughter on her 21st
birthday, how her voice had grown
around her like the rings of a tree.
As soon as she recognized my voice,
she hung up. A good father blames himself.
I checked every centimeter of the satellite,
inside and out, polished her again
and again until I could see the glow
of my cigarette from across the room
on her silver. She was a perfect sphere
ready to catch the sun and share it
with the world through her radio—
she didn't see the USSR, the USA,
no space race, just space, this speck
swaddled in blue, green, and brown—
freckled with city light. Hearing Sputnik
on the radio after her first orbit was
like when your child says your name
for the first time; you listen to a place
that wouldn't have existed without you.

The Parable of St. Francis and Tetra, the First Cloned Monkey

After his crusade against his neighbors in Perugia, his capture and imprisonment, St. Francis became two men. One terrified of what he had done, another taken up in the raptures of who he was becoming close to. Both lived within him and like nuclear fusion mushrooming a miniature sun; this tension radiated his love for nature and sanctified him.

Half the saint has always thought himself unworthy, while the other half had gone beyond, permeating the world like hot water around a tea bag, seeping out the flavor and color that revives as it enters another body.

Tetra was split in half when she was eight cells old to become two genetically identical sacks of embryos.

St. Francis wants to be like Christ, two natures in one, but he cannot talk to Christ, the echo within the silence living within. So he asks the monkey, Tetra: do you feel the gap of being half of another? Or do you feel as if the two you became is better off with each other in this double life?

Tetra, looking like every other rhesus macaque to those who do not have eyes to see, says, *I am a simian among simians; there is no truer self than that for us, cousin sapien.*

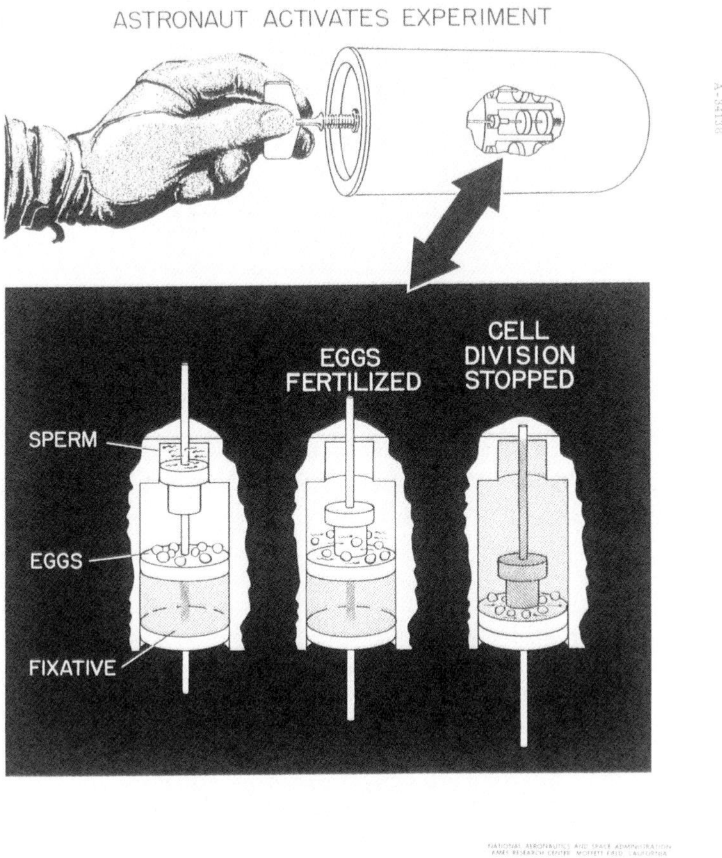

MAKING LIFE IN THE VACUUM

Father Nescio Buying Baby X

Like every other ovulator, my eggs
were harvested at birth and saved
in the national bank, which got its sperm
from the holographic brothels that offered
it up like market fish. Geneticists ran models to
determine which combinations would create
superior offspring, CRISPRing away
disability or susceptibility to disease,
while psychologists analyzed potential
parenting candidates. Buyers like me,

lucky among the waiting. At the guardian
school, our cohort discussed between classes
what features we'd chosen, eye and hair
color mostly, while the fetuses grew inside
plastic wombs. We compared the monthly
photos—my bird of paradise hatching inside
gelatin, growing into its cooing habitat,
another's kicking at the snaking coaxial

cord. I could hardly believe when the drone
delivered my immaculate babe
and supply box. I signed the lease
the drone held, then took the swaddled infant
into its new habitat. The future
pooling into its beady eyes that pulled
the world apart, bit by bit, into itself
as I wondered what name came with it.

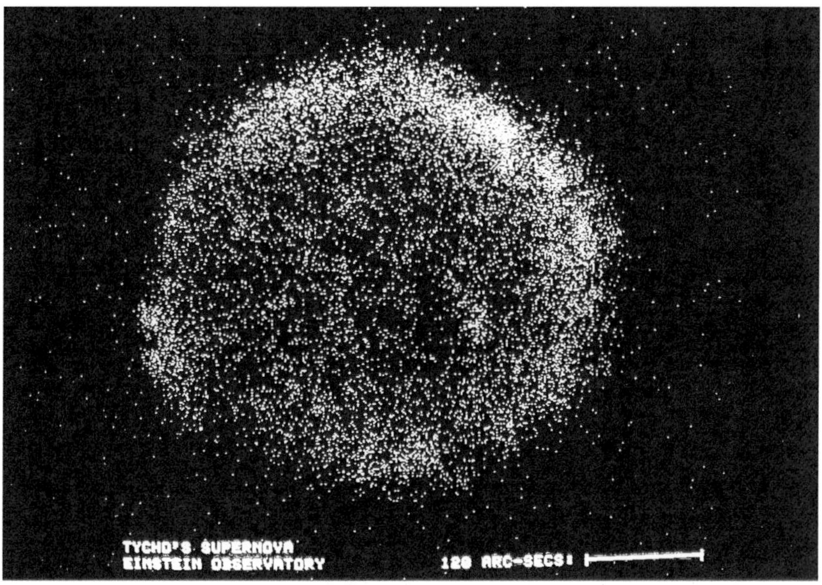

X-RAY OF BIRTH

Elliptical Hymn to Saturn

Jewel of the night crowned with an apatite
aurora, your hoary hair roars over
the black bathtub of the Lord as you float,
light as a grain of sand rippling an hour
to the brim of time. Your sickled moon rings
the meal bell, salt dissolving on your lips—
Prometheus plows a division like a harpist
striking a string into a haunted chord,

the moan of gods eaten by time. We feed
upon the plunder of the eye, the telescope
stocking the silos for more of your sky—
the clouds of peanut butter spread on toast,
gales of wind on seas of oat and rye, waves
bowing to the invisible presence whose
sacrament is bread and grapes dying
to be reborn in time as the body, a vessel

fermenting the firmament, thirsting for the Nile
flowing upstream, flooding the banks
with so much grain, only a scribe could say
how much is there to share. The only way
to save what's been consumed is to reap
again. Though no bees fly from your honeycomb
storm to pollenate flowering asteroids
for seeds to fill your tilled rings, you're the fruit

in the garland of the spheres, the manna
in the desert vanishing each morning,
the nectar that the hummingbird flies
thousands of miles from its home for.

Father Nescio's New Eyes

They gave my eyes extra cones and rods
to see the same colors as the lobster
scuttling after coral eggs in the dark. I saw

so much. Yet when I looked at a Dutch
masterpiece, all the layers of paint came
through at once, grayed like paint thinner.

They tried to improve beauty by upgrading
the eye of the beholder. And the rainbow's
promise became an atmospheric flood.

FIFTH FRAGMENT OF THE HYPATIA STONE

nature thrusts

the deep

up
on

us

and we
rarely

believe

our
selves
ready

to
weigh

the
split and
split and
split

again

avoirdupois

night

URANUS,
THE CARTOGRAPHER

FATHER NESCIO'S SIXTH PRAYER

Bit by bit, I recover myself
from swallowing the burning coal
the seraphim held out.

It is astonishing how long
the brain held over the flame

lasts. I plunged one of his feathers
down my throat to vomit

the ember out. Now I can say
I've seen the day of the Lord:

not in the solar wind that bends
magnetic fields, nor in

the fire that burns fire into plasma,
but in the whisper

of the marioplex of suns
left to express presence.

Let me travel far from this sun,
the only sun I've ever known,

to another of your blinding faces.

FAUST WATCHING APOLLO 11

- Wernher von Braun, 1969

Like the Statue of Liberty leaving for the moon,
Apollo honed a cone-tipped pillar of flame,
breaking with sheer Samsonesque strength
the gravity of Earth. Then, in the slow Nile of space,
it floated, a metal basket wading in black water.
When we were told to go forth and multiply,
it was a promise: the tree of our species would spread
beyond this island world. We have more right to land
on the moon than a comet— we, a helicopter
seed in the wind bringing life to the lifeless.
Seeing Earthrise over the lunar horizon was the only
horizon out of all those yet to come that I would see.
I was like Moses, not allowed to step foot in the
promised land, only able to see it from a distance.
But to have seen our home as another world, a
solar satellite in the endless night: it was something.
I built the ladder to the heavens I would never
climb, unless I hung myself on the tree of life—
walking on the regolith dust, each immovable print
as good as mine as anyone else's. Odin lost an eye
to gain wisdom, I lost my chance to see the moon
for myself, but I dredged up future: the world
given to us so that we could be given to the cosmos.

The Parable of St. Francis and the First Monkeys in Space

They strapped Miss Able into a tiny metal jacket, placed her right hand across her heart as if she were pledging her allegiance. Set her chin so she sat looking up— a mystic Rhesus monkey beholding the giver of bananas. She didn't go alone. With her was a hand-sized squirrel monkey, enclosed in her shoe box case with chamois strappings, her name spelled on a plastic beaded necklace, B-A-K-E-R, her tail puffed beside her like a miniature bushy palm tree. Six minutes after launch, they enjoyed a few minutes of weightlessness before falling like fruit back to earth.

For so long, St. Francis longed to enter the divine heights with someone so that he could share that joy in a common tongue. But he went alone and came back to find that animals were the better listeners. St. Francis asks Miss Able and Miss Baker if, as they orbited the Earth, becoming a celestial garland, they felt a greater kinship to the creatures of the Earth because they had each other.

Miss Baker, who is grooming Miss Able, tells the saint, *we monkeys are a barrel. When a fruit falls in the forest, we know it by the sound of other monkeys gathering.*

MISS UNIVERSE

Father Nescio and the Algorithm

There were no more ready-mades
—generations of consumer info
on hand, made every item custom, printable on
demand. We were paying to eat a slice
of our own bespoke-minds.

Though we were billions upon billions
swarming
across planets and moons, each of us

was known, precious as a protozoa held
under the microscope's eye. The very
hairs of my heads counted, my every move
tailored from within the womb.

In the past people were sick as their secrets,
now our cravings have been fine-tuned as
                                        digital watches.

ELLIPTICAL HYMN TO URANUS

So if there is a journey pray that it be long
-Zbigniew Herbert

Other blue dot, standing upright, spinning
a mohawk of yo-yo rings around your clouds,
be the key opening me up to the map
precursing scripture. To the black shrouds
disguised by how we connect the starry dots.
On the Nebra Sky Disk your teal methane
chews gas to diamond while gold rivets
of the Pleiades show the time to sow grain.

The planets triangulate and a gold ship pulls
us to spring. We no longer thank the sky
for fields of gold, shelves stocked with livestock—
cost controls our loss. But when lost, we cry
out the names of those we love, and if no one
comes, we call for anyone. Is anyone
out there? we ask with radio telescopes
in the desert, bored with static, rerunning

scans for transmission. Tonight, life bobs
and wades as we enter the hours that oil
the shore. The red flares we launch are letters
in a bottle, not lifebuoys. We've mapped
ourselves on the Pioneer plaque naked and
waving and placed among fourteen pulsars
closest to our sun. Above us floats a tiny threat—
the anatomy of an atom. We're a lonely,

fearful species, reaching up to meet you,
dear otherworldly. Use this map to help us,
find us, show us how the night doesn't end
the day, that searching was half the answer.

Sixth Fragment of the Hypatia Stone

pierce my side

spill the fat

of my distant
system

up
on your
cutting
block

I've
sailed

so far
to be

a window
of
where

you're going

an inch of the expanse

NEPTUNE,
THE OCEANOGRAPHER

Father Nescio's Seventh Prayer

Teeth Mother, you make the walls
drip like the resinous tide ebbing out

an oil spill. The flame inside
the yak skull animating all past

lives leading to that Cosmic Body.

Pre-chew my organs, spit feed me my inner ooze,
pass on the enzymes and bacteria
I need to digest like the sun burning on itself.

With you, Teeth Mother,
of whom shall I be afraid? Thou art
the breast milk of stars, able to smother
my enemies like tadpoles raised
between two fingers. Through you I'm given

to all parts. Dark matter, dark energy,
the ninety-seven percent of the empty that
consumes me.

The white horns of plenty grow
behind my head. I rail and rise

through the gates of horn into the fields—

The horde of corpses sent
from the dump outside Jerusalem,
I trample underfoot.

Their sticky fingers trying to cling
through the muck
won't smear a single hoof.

The Parable of St. Francis and ANDi, the First Transgenic Monkey/Jellyfish

The gold nimbuses in medieval frescos were how painters depicted saints who were so open to God, God showed through them. Like planets and moons reflecting the sun, the closer you orbit a center, the more light you radiate.

The heartless, brainless jellyfish that can potentially live forever by splitting into copies of themselves have become a part of ANDi, and ANDi a part of that first jelly, numinous in its singularity. Christ, St. Francis thinks, is also an incarnation of two. A merger of matter and mind, spirit and law, flesh and selflessness.

St. Francis asks ANDi if being the blending of two species has given him insight into the mystery of the incarnation. *The gods who made me thought I'd glow,* he says, disconcerted, *but only my still-born brothers did.* Then ANDi asks, *did death make you radiant?*

St. Francis thinks about the fields of slain crusaders, their armor glinting in waves of heat, and says he doesn't understand the mysteries of light but that everything he sees in the world he sees by it. The monkey looks at the glowing umbrella of St. Francis's nimbus and sees in him a distant relative pulsing through the sea, the moon broken into moons on the waves.

ELLIPTICAL HYMN TO NEPTUNE

Even with all the muses behind me.
- Wisława Szymborska

I'm a reflection of the sea seeing you,
O god of the deep. My short Odyssey
traces its umbilicus to the Marianna trench,
where life is star-like, a cast-down burning
lake aflame. In my spine floats a tail
feather, a sheathed snake scale, a fish darting
among the nautili curled in their golden
means. We found you by math, trailing

waves of undulating gravity, tracking
Uranus's anomalous orbit to your teal
chariot—neuron to neuron, thought moves
through the dark methane that peels
a diamond sea-storm howl, cracking the
whip on the horses of your wind, driving them
supersonic. When I hold a seashell
to my ear, your misty rings and moon thrum—

Nereid's icebergs creak, Triton's geysers
breach. We return to the sea like a turtle
to its breeding shore, remembering how good
it feels to be small and searching, but I've come
to you unhoused and hopeful, seeing the city
light on the shore and mistaking it for the
moon. Space is more ferocious than a shark
bite when it grins its endless teeth—piercing

constellations that guide us while we voyage for
that familiar deep. From your harbor, we see our
own blue dot with the only life we know: lightning,
a calm, and us as the amino acids in a wine-dark sea.

NEPTUNE'S HORSES

Finally, I Am Sergei Korolev Again

- 1966

When I opened the heavens, my name was taken—
as Saul became Paul, as Joseph became Stalin,
and my people, comrades, so I became *Chief*
Designer. Ordained to reach beyond my own
hand and touch the stars for my people, hungry
as termites crawling on the ceiling of cathedral
factories. I whisked us off the imagined flat world
of the map, erasing the pencil-drawn new order,
and delivered an uncheckered planet. My rocket
was more than a distant view of a small rock, wet
with moss. It was a truth incarnate— a transcendence
into the womb of being, a vision of how we belong.
The needle inside the inflated balloon, rising. Popping
hemispheres and showing the mind as it is: green
grooves pushing against themselves into folds that slosh
in salty fluid. When we turn away from the sun, our
cities are the constellations saying what they mean.
When I see my rocket, I know the interior is the
longest horizon, the endless sun of summer,
the endless night of the Arctic winter over that
archipelago of what we must say and do to stay
alive. What we keep to ourselves is our soul, no
matter how insubstantial. What we give to each
other is the presence. When I died, they returned my
name to a person who was no longer there—I belong
now to my people, and my name has become
a word for more than the man who wasn't anyone's
comrade, anonymous on the streets of Kiev.

THE INTERNATIONAL SPACE STATION
OVER UKRAINE

FATHER NESCIO AND THE LAST WINTER

The astronauts waking from their cryogenic sleep,
frost amidst their body hair.

Seventh Fragment of the Hypatia Stone

lean
close

to my
broach

sarcophagus

hear

the oneiric

opuscule

of my
heart

Elliptical Elegy for Pluto

I am not there either.
- Ai

Death, you who are and are not a planet
with an icy heart, ferry me on the tightrope
of the Kuiper belt so I may toe the heliopause
like a slip knot that can't escape the rope.
Place your minted coins under my tongue
so that I may speak something true of endings.
When the dim light of the sun slides along
your glacial surface, a house finch flies

the blue eggshell of atmosphere to its own end.
I want to drink from your blue oblivion,
bury a satellite to return to the signal.
Have mercy on me, I'm clay from a far
away oven. Who I am is linked with who
has left. Look down at my hands, see
the cave fire that guards my ancestors against
the night's invisible raids. My palm lines

spear the mammoth painted on the walls,
resurrecting the eaten animal in the flickering
light. We've always wanted the dead to crawl
again. If there is a soul it's a shared, uncleared
transit. When Clyde Tombaugh found you
orbiting a gravity well with Charon, he thought
he was uncovering the ninth muse the music
says must be there,

balancing the planets. Though it was an elysian
field of undead spheres, when New Horizons
flew by you with one ounce of his ashes, both
of you were the first of us to be and not be risen.

Father Nescio and The Ferry

I paid the boatman,
sliding the holographic coins
into his blue purse, a stream of silver fish.
Up we go, said the boatman,
closing the door on another sphere,
a moment of darkness then the blue lights
opening overhead. Most people took a seat
on one of the benches, but I stood near one
of the windows. My planet was green and
red on the surface but brown as tigers-
eye the higher we went. The green algae
rivers and reclamation lakes and the scrub
brush and pine forests mixed
with the gold of city lights
and the ever red, red regolith,
a bead fit for a god-king.

It takes so little time to leave a world
behind, you'd be forgiven
for thinking you could return to it
just as quickly.

ACKNOWLEDGMENTS

Many thanks to the editors of the following magazines in which the following poems appeared:

Atticus Review: "Luxury Gulag," "Rock-a-Bye, Sputnik"
Aquifer: *The Florida Review Online*: "Tomorrow Land," and "Man's Dominion"
Cider Press Review: "The Devil in the Desert," "The Coldest Prison, "—special thanks John Bradley for anthologizing the latter poem in *And Blue Will Rise Over Yellow: An International Poetry Anthology for Ukraine* (Kallisto Gaia Press, 2023).
Heavy Feather Review: "Father Nescio and Virtual Fantasy Machine," "Father Nescio and the Forgiveness of Machines," "Father Nescio And the Ark of the Future," "The Second Prayer of Father Nescio," "Father Nescio's Exorcism."
Permafrost: "Faust Watching Apollo 11"
Rabid Oak: "Elliptical Hymn to the Moon," "Elliptical Hymn to the Neptune," "Elliptical Hymn to Pluto,"
White Stag: "Elliptical Hymn to Venus," Elliptical Hymn to Jupiter"

***

Thank you to my teachers, David St. John, Beckian Fritz Goldberg, Alberto Rios, Jeannine Savard, David Wojahn, Kathleen Graber, Greg Donovan, Harrison Candelaria Fletcher, Liz Canfield, John Freyer, Josh Rathkamp and Norman Dubie. No student can thank their teachers enough.

To my Peace Corps community and my Solar System Ambassador community for giving me opportunities to prove the world has no borders. To NASA and the Smithsonian Institute for letting me use their images.

To the Jungian alchemist Gil Stafford for helping me see into and through my shadows.

To Amadou Guindo and Teatr Baraka for letting me read some of these first poems in Krakow, Poland; likewise, to Nicanor and Fernanda Escalera, who invited me to read my poems in their beautiful bookstore, Más Allá, in Puebla, Mexico.

To Adam Zagajewski for tea and espresso in Krakow and your indispensable wisdom in helping me balance the poems and their echoes.

To everyone who read earlier versions of poems in this book and encouraged me to keep writing: Jordan Dahlen, Justin Krider, and David Moakley; and to those who read the entire manuscript and provided indispensable feedback: Justin Beloit, Hugh Martin, and Michael Shewmaker. To those who humbled me with their blurbs: Dan Beachy-Quick, Dexter Booth, Hugh Martin, Michael Shewmaker, Kim Vodicka, and Frank White.

To Courtney and Tucker Jameson for believing in this book and welcoming me into the White Stag herd.

To my parents for their love and religiously watching Star Trek as we grew up. To my sister for her wisdom. To my mother-in-law for her help in taking care of my son. To my son for opening a new dimension in the universe. To my cat Zbigniew Herbert the IX for the purrs and the kneading (even though you rip holes in my pants and cut my skin—you're too damn cute to move) and often being my first listener.

Above all, thank you to my wife, Victoria Miluch, for opening me to the world and bearing the cosmos; for reading, re-reading, and making brilliant suggestions. I cannot imagine any word or world without you.

IMAGES

- **Cover image:** Photo by Long, Eric. "Monkey, Able." Smithsonian Institute.
- **All astronomical symbols:** NASA.
- **Figure 1** (pg. 3): NASA. Spectroheliogram of the solar eruption of June 10, 1973.
- **Figure 2** (pg. 5): NASA. Drawing of proposed Pig Capsule in Little Joe capsule. 1963.
- **Figure 3** (pg. 17): Left image: Public domain. Photo of Félicette. Right image: NASA. Drawing of the A-4 rocket. The French rocket that sent Félicette to space had its roots in this Nazi rocket. 1940.
- **Figure 4** (pg. 21): NASA. Apollo Configuration of Saturn Model in the 8X6 foot supersonic wind tunnel. 1961.
- **Figure 5** (pg. 28): NASA. Early rockets. 1940.
- **Figure 6** (pg. 35): NASA. Food for the Gemini 7 space flight. 1965.
- **Figure 7** (pg. 40): NASA. Viking Lander 2's soil sampler collector arm pushes a rock on the surface of Mars. 1977.
- **Figure 8** (pg. 46): NASA. First evidence of a ring around Jupiter; photo taken by Voyager 1. 1979.
- **Figure 9** (pg. 55): NASA. Diagram of an experiment to test the effects of zero gravity on living cells for the 1965 Gemini-Titan 3 mission.
- **Figure 10** (pg. 57): NASA. Supernova in Cassiopeia constellation. 1980.
- **Figure 11** (pg. 65): NASA. Miss Baker with a model rocket. 1959.
- **Figure 12** (pg. 73): NASA. Voyager 2 photo of Neptune's rings. 1999.
- **Figure 13** (pg. 75): NASA. View of the ISS over Earth. 2000.

WHITE STAG